BUILDING BLOCKS OF COMPUTER SCIENCE

HARDWARE

Written by Echo Elise González

Illustrated by Graham Ross

a Scott Fetzer company
Chicago

World Book, Inc.
180 North LaSalle Street
Suite 900
Chicago, Illinois 60601
USA

For information about other World Book publications, visit our website at www.worldbook.com or call 1-800-WORLDBK (967-5325).
For information about sales to schools and libraries, call 1-800-975-3250 (United States), or 1-800-837-5365 (Canada).

© 2021 World Book, Inc. All rights reserved. This volume may not be reproduced in whole or in part in any form without prior written permission from the publisher.

WORLD BOOK and the GLOBE DEVICE are registered trademarks or trademarks of World Book, Inc.

Library of Congress Cataloging-in-Publication Data for this volume has been applied for.
Building Blocks of Computer Science
ISBN: 978-0-7166-2883-5 (set, hc.)

Hardware
ISBN: 978-0-7166-2884-2 (hc.)

Also available as:
ISBN: 978-0-7166-2892-7 (e-book)

Printed in China by RR Donnelley, Guangdong Province
2nd printing August 2021

STAFF

Executive Committee
President: Geoff Broderick
Vice President, Finance: Donald D. Keller
Vice President, Marketing: Jean Lin
Vice President, International Sales: Maksim Rutenberg
Vice President, Technology: Jason Dole
Director, Editorial: Tom Evans
Director, Human Resources: Bev Ecker

Editorial
Manager, New Content: Jeff De La Rosa
Writer: Echo Elise González
Proofreader: Nathalie Strassheim

Digital
Director, Digital Product Development: Erika Meller
Digital Product Manager: Jon Wills

Graphics and Design
Sr. Visual Communications Designer: Melanie Bender
Coordinator, Design Development and Production: Brenda B. Tropinski
Sr. Web Designer/Digital Media Developer: Matt Carrington

Acknowledgments:
Art by Graham Ross/The Bright Agency
Series reviewed by Peter Jang/Actualize Coding Bootcamp

TABLE OF CONTENTS

How does a computer work? 4

So many computers! 6

External hardware 8

Internal hardware 12

Chips 14

CPU 16

The motherboard 18

Memory 20

Expansion cards 24

Hardware and software
 work together 26

Glossary 30

Go online 31

Index 32

There is a glossary on page 30. Terms defined in the glossary are in type **that looks like this** on their first appearance.

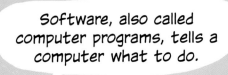

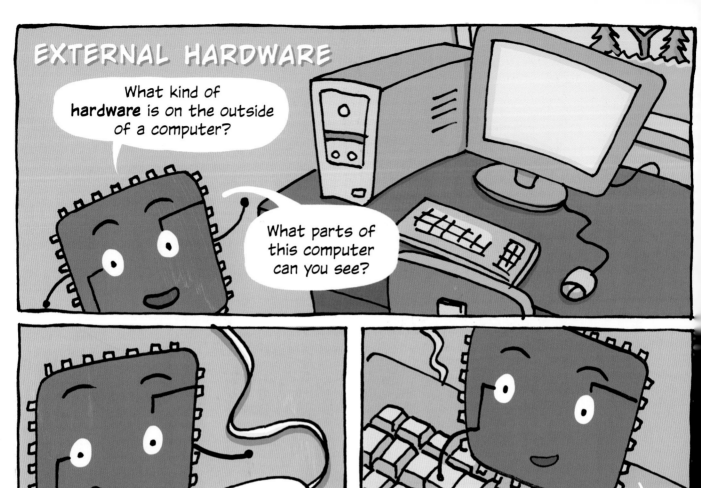

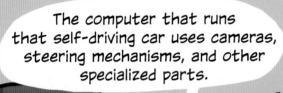

CHIPS

Hey, guys!

These are my buddies, the other **computer chips**.

Some of us are **processors**, and some are **memory chips**.

Processors carry out the instructions that make up computer programs.

They perform important operations for the computer.

Memory chips store **data** and computer programs.

They're responsible for keeping track of much of the computer's information.

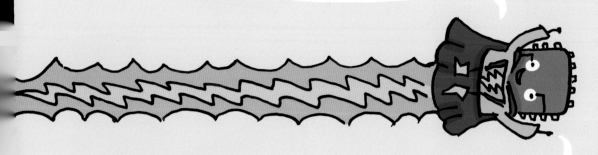

In this way, they can share information with the rest of the system.

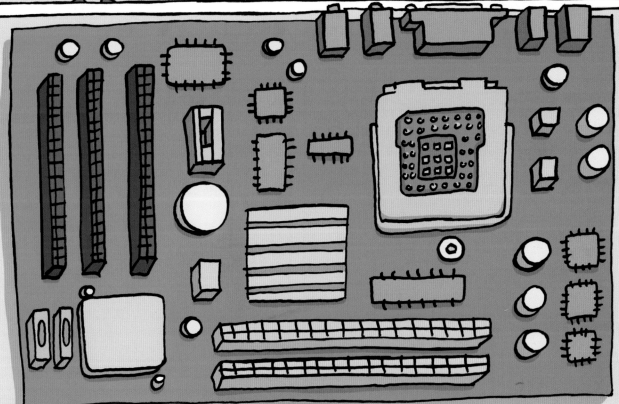

The motherboard also has many bundles of circuits called **chipsets**.

The **CPU** uses these chipsets to send instructions to the rest of the hardware.

EXPANSION CARDS

The slots you see here on the **motherboard** are called expansion slots.

They hold the **expansion cards.**

The expansion cards have a special job—to enhance the computer's performance.

The **network card** lets the computer connect to the internet and other computer networks.

HARDWARE AND SOFTWARE WORK TOGETHER

GLOSSARY

algorithm a set of step-by-step instructions used to write computer programs. Algorithms are also used to solve math problems and other problems.

binary digit a 0 or 1. These are the two digits that make up machine language.

chipset a collection of computer chips that helps data flow between the CPU and other computer parts.

circuit a loop that an electric current can follow.

computer chip a tiny piece of silicon that holds an electronic circuit. A circuit is a pathway that an electric current can follow.

CPU (central processing unit) the main microprocessor of a computer. It sends instructions to many other computer parts.

data information that a computer processes or stores.

expansion card a circuit board that can be connected to a motherboard to add extra capabilities to the computer. An expansion card can enhance the visual, audio, or other capabilities of a computer.

external hardware the parts that make up the outside of a computer or device.

external storage device a device used to store memory outside of a computer. A USB stick and a DVD are two examples of external storage devices.

hard drive a device that stores data on a disc or magnetic platter.

hardware the physical parts that make up computers and other electronics.

heat sink a material or device that removes heat from a piece of hardware.

input device a device that sends information from outside a computer into the computer. Computer microphones, keyboards, and video game controllers are examples of input devices.

internal hardware the parts that make up the inside of a computer or other electronic device.

machine language the code used to communicate programs to a computer's hardware. Machine language is made up of binary digits.

memory chip a kind of computer chip that stores information.

microchip a very small computer chip.

monitor a screen that outputs visual information from a computer.

motherboard a board that holds circuits, the CPU, and various computer chips inside a computer.

output device a device that sends information from inside a computer to outside the computer. Monitors and computer speakers are two examples of output devices.

power supply a piece of hardware that provides other pieces of hardware with electricity.

processor a kind of computer chip that performs calculations for the computer.

RAM (random-access memory) a kind of computer memory that temporarily stores data while the computer is being used.

ROM (read-only memory) a kind of computer memory that stores data long term.

semiconductor a material that can conduct electricity better than *insulator* materials, but not as well as *conductor* materials.

silicon a material used to conduct electricity in computer chips.

software computer programs.

solid state drive (SSD) a device that stores data on a microchip instead of on a platter.

GO ONLINE

Now that you know all about the hardware that makes up a computer, would you like to see if you have what it takes to build one? Go to this website and click on the Build a Computer activity! You'll also find a Hardware Memory Game, a Make Your Own Circuit activity, and many more fun activities!

www.worldbook.com
/BuildingBlocks

INDEX

algorithm, 5, 27

binary digits, 5

case, 8
central processing unit (CPU), 16-17, 27
charging cable, 9
chipsets, 19
circuits, 19
circuit board, 18
code, 27
computer chip, 4, 14-15
computer program, 14, 26-28

data, 11, 14, 20, 22-23

expansion cards, 24-25
external hardware, 8-9, 15
external storage device, 9

fans, 13

graphics card, 25

hard drive, 22-23
heat sink, 17

input device, 10-11
internal hardware, 8, 12-13, 15

joystick, 10, 17

keyboard, 8, 17

machine language, 5
memory chips, 14
microchip, 23
monitor, 8-10
motherboard, 18-19, 24-25
mouse, 4, 8

network card, 24-25

output device, 10-11

platter, 22-23
power supply, 13
processors, 14, 23

random-access memory (RAM), 20
read-only memory (ROM), 20-21

self-driving car, 9
semiconductor, 15
silicon, 15
software, 4-5, 26-27
solid state drive (SSD), 23
sound card, 25

video game console, 9

Children's 004 GON
González, Echo Elise
Hardware

01/27/22